Lots of Fun Number Tracing Practice!

by Handwriting Time

Lots of Fun Number Tracing Practice!
by Handwriting Time

The images of the numbers 0 to 20 used on the cover and interior of this book are derived from two images (copyright katarzyna b), which were purchased from Adobe Stock with an Extended license. This license permits an image to be used on the cover and interior of a commercially available physical book where the image purchased forms a significant proportion of the value of the book.

The bird images used on the cover and throughout this book, are derived from a single image (copyright hibousunart) which was purchased from Adobe Stock with an Extended license. This license permits an image to be used on the cover and interior of a commercially available physical book where the image purchased forms a significant proportion of the value of the book.

The tree background images used throughout part 2 of this book were derived from an image (copyright blueringmedia), which was purchased from Adobe Stock with a Standard License, which permits the image to be used in a product for resale providing the main value of the product is not the image itself.

The image of a hand holding a pencil, used on the cover and in the interior of this book (copyright luchschen) was purchased from iStockphoto.com with a Standard License. This license permits an image to be used on the cover and interior of a commercially available physical book.

The font used in naming the numbers on the cover and interior of this book is Simple Print. This font is available under the SIL Open Font License Version 1.1 which permits the font to be used on the cover and interior of a commercially available physical book.

The font used on the cover and interior of this book to illustrate the traceable numbers on each number page is Print Clearly, copyright © 2009 Blue Vinyl Fonts. This is a Freeware Font which permits the font to be used in a commercial manner. The font used on the title and copyright pages of this book is FreeSans. This font is licensed as GPL (GNU General Public License) which permits a font to be used on the cover and interior of a commercially available physical book.

The font used for the title and age recommendation on the front cover is One Stroke Script.* The font used on the Part 1 and Part 2 title pages, and throughout Part 2 in the interior of this book is Segoe Print.* The font used on the cover and interior of this book to number the arrows indicating how each letter should be formed is Trebuchet MS.*

*Desktop Licenses for these fonts were purchased from Fonts.com. A Desktop License permits the use of a font on the cover and interior of commercially available physical books.

Disclaimer: If this book is used as an educational aid/tool, it is up to you (the user, parent, guardian, teacher etc) to confirm that this book is useful, and you take full responsibility for the use of this book. In no event shall the author and/or publisher be liable for any damages whatsoever arising out of the use of or inability to use this book.

ISBN-13: 978-1546346340
ISBN-10: 1546346341

Part 1: Tracing numbers

Firstly, follow the gray lines
to form the numbers.

Then, try writing the numbers
without the gray lines.

zero

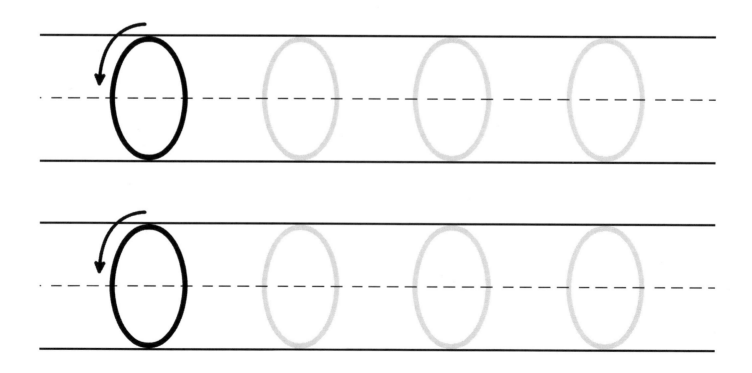

one

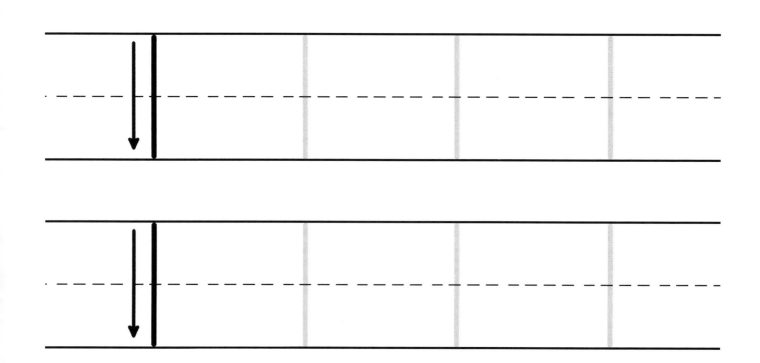

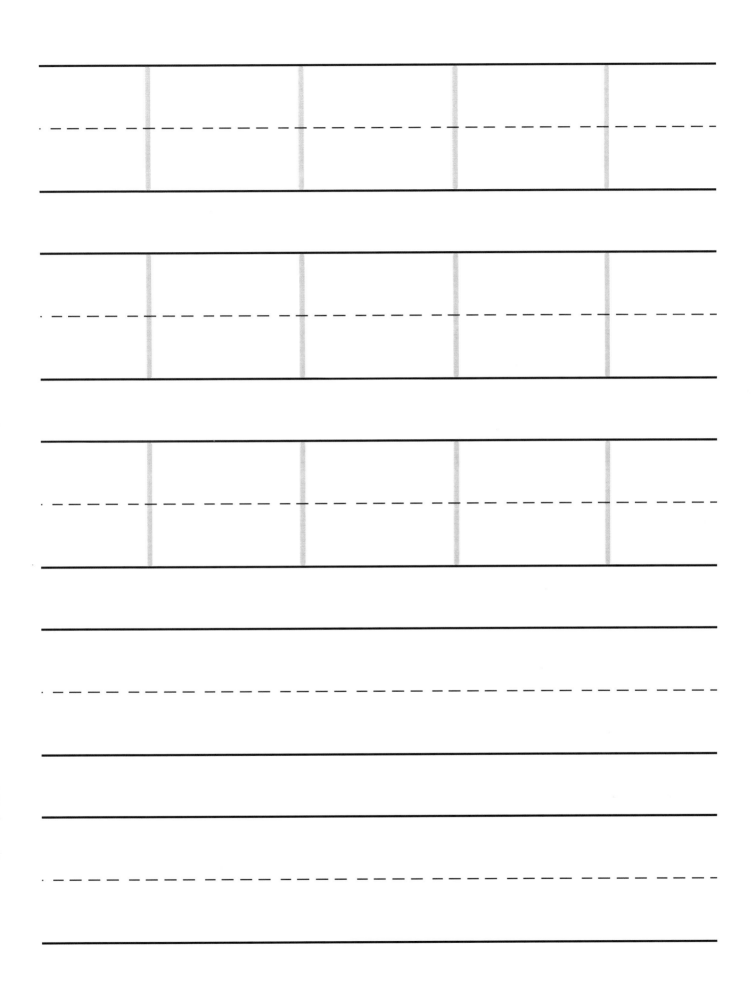

two

2 2 2 2

2 2 2 2

2 2 2 2

three

3 3 3 3

3 3 3 3

3 3 3 3

four

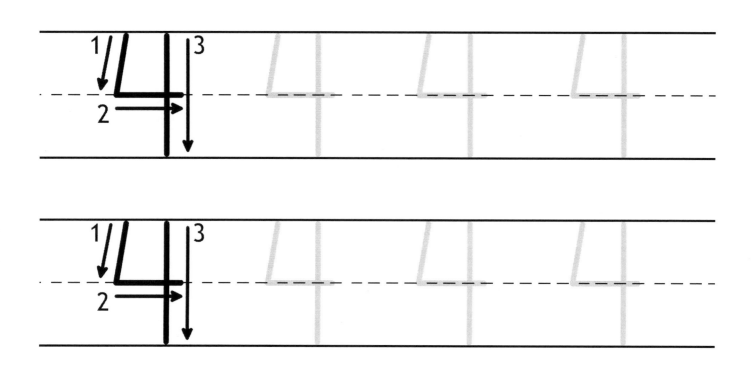

five

5

5 5 5 5

5 5 5 5

5 5 5 5

six

6 6 6 6

6 6 6 6

6 6 6 6

seven

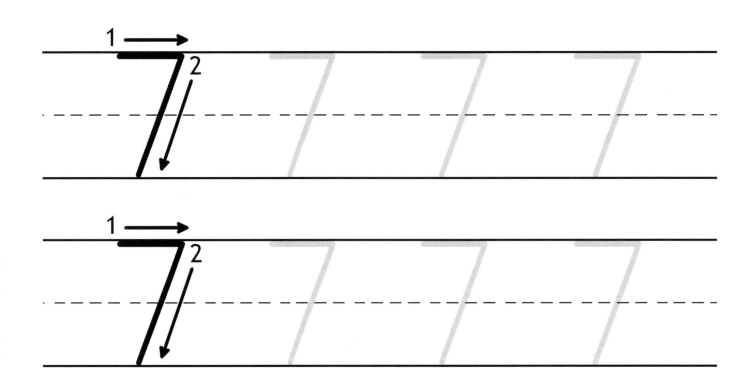

eight

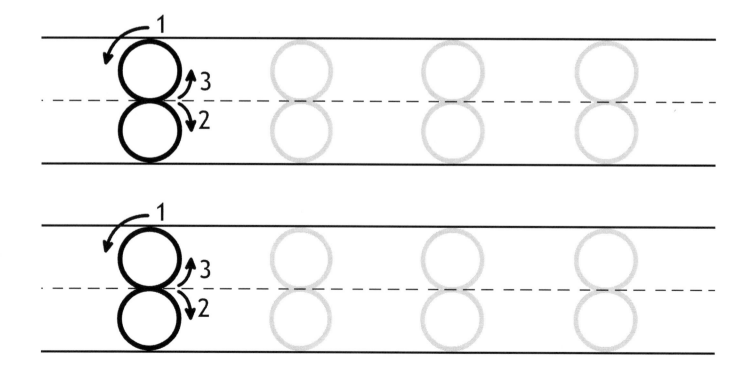

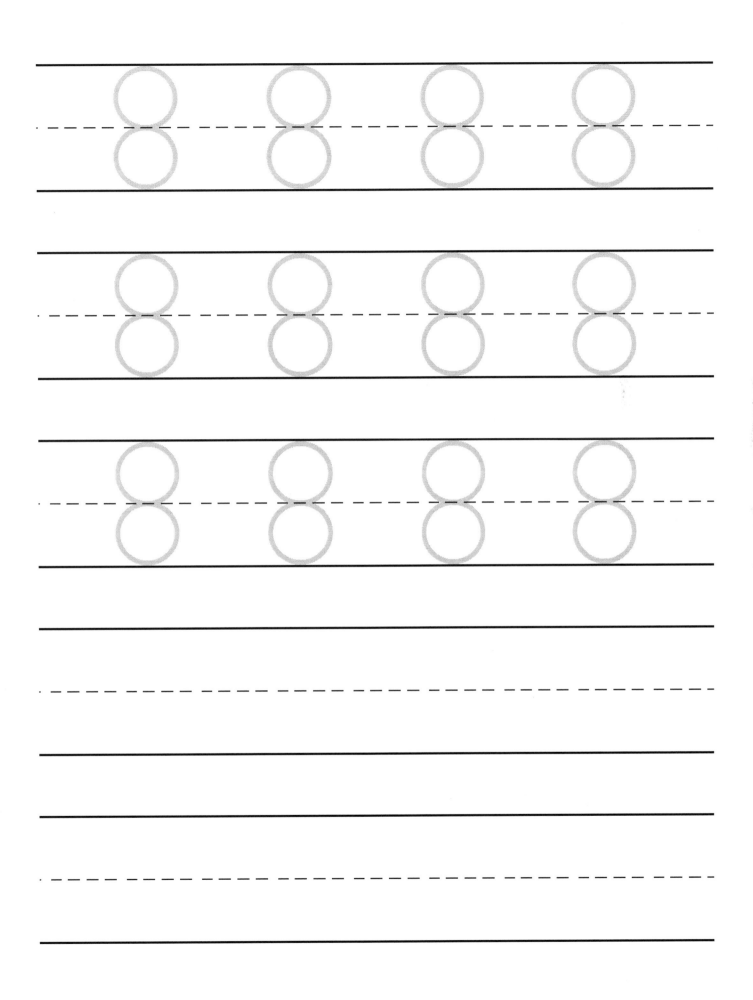

nine

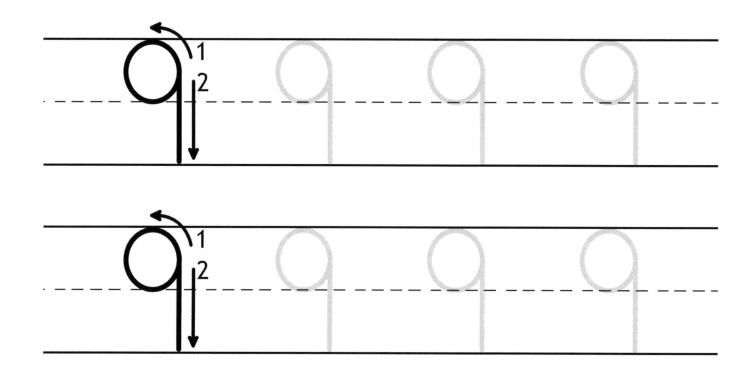

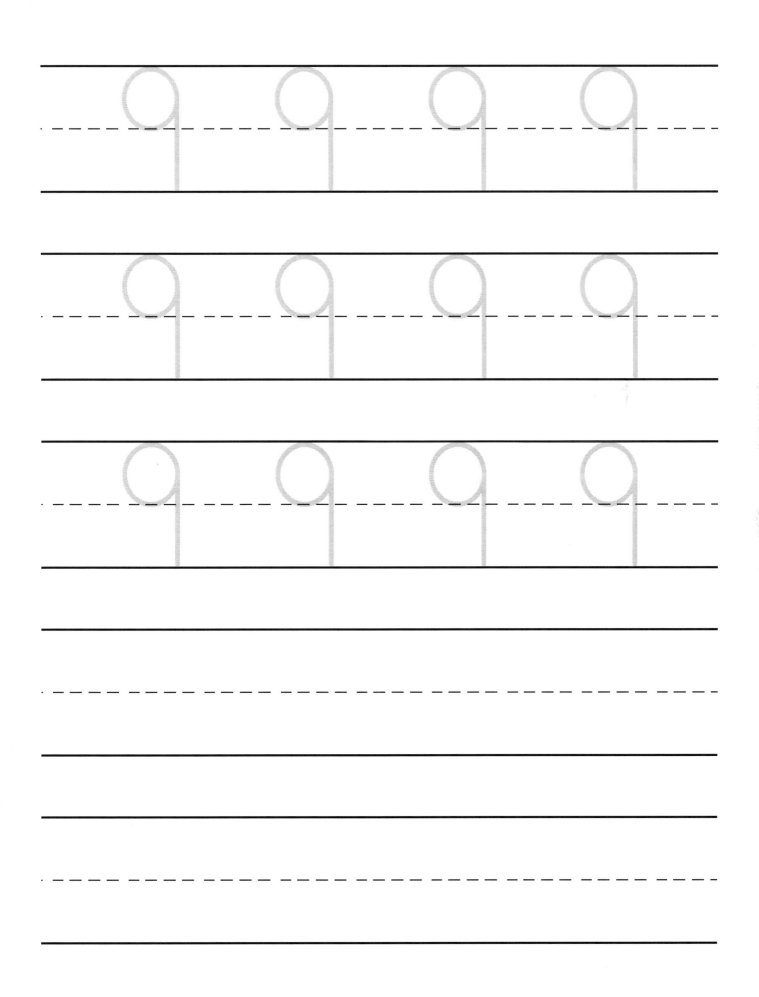

ten

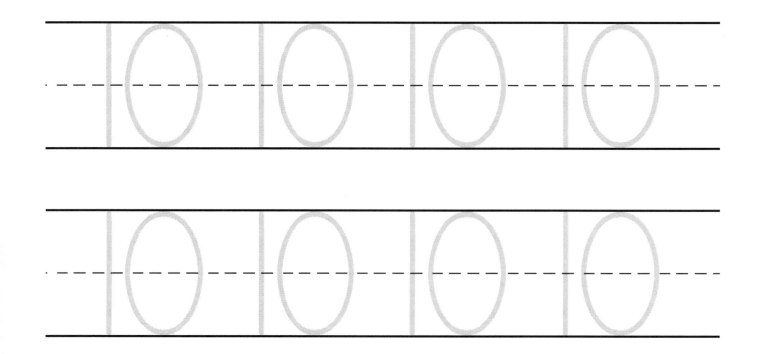

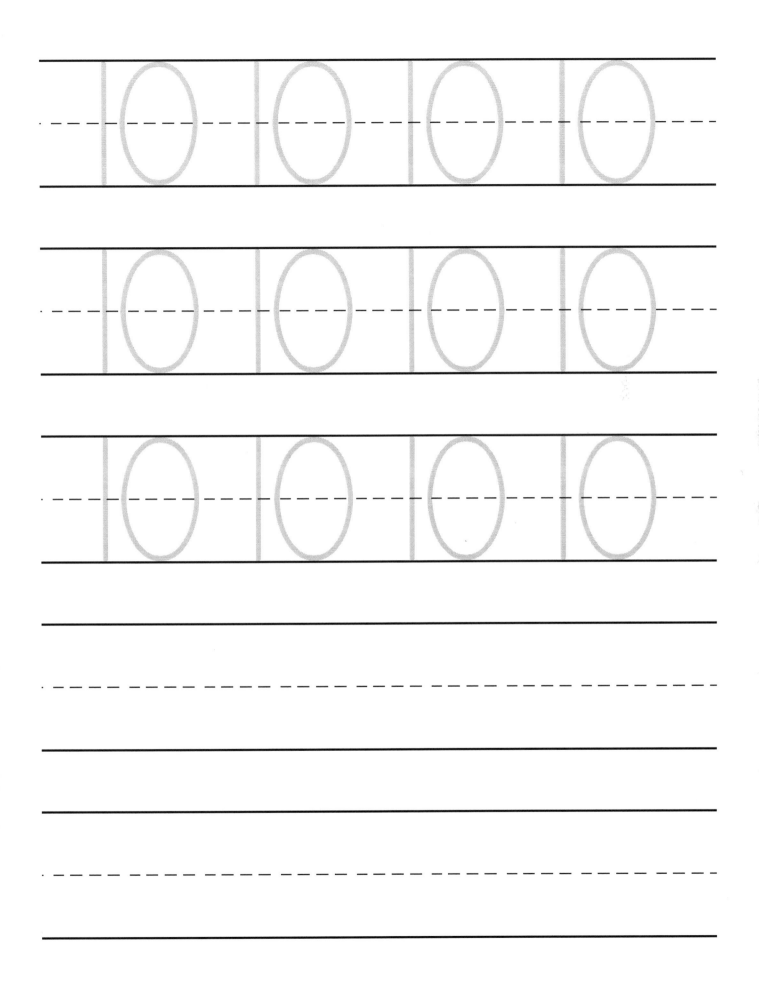

eleven

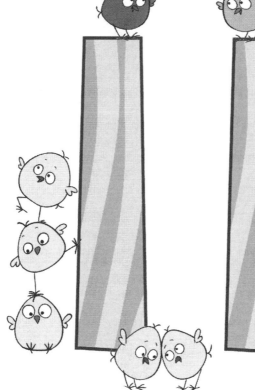

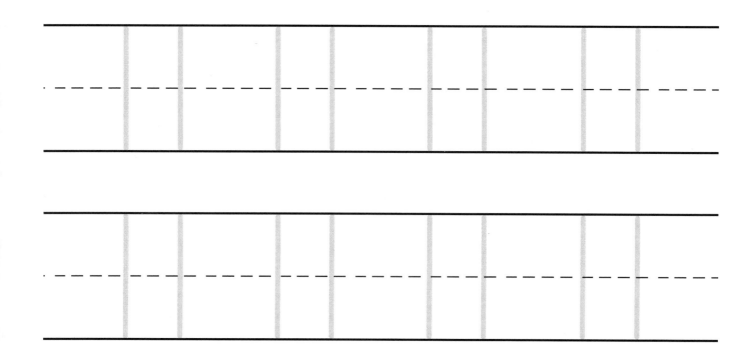

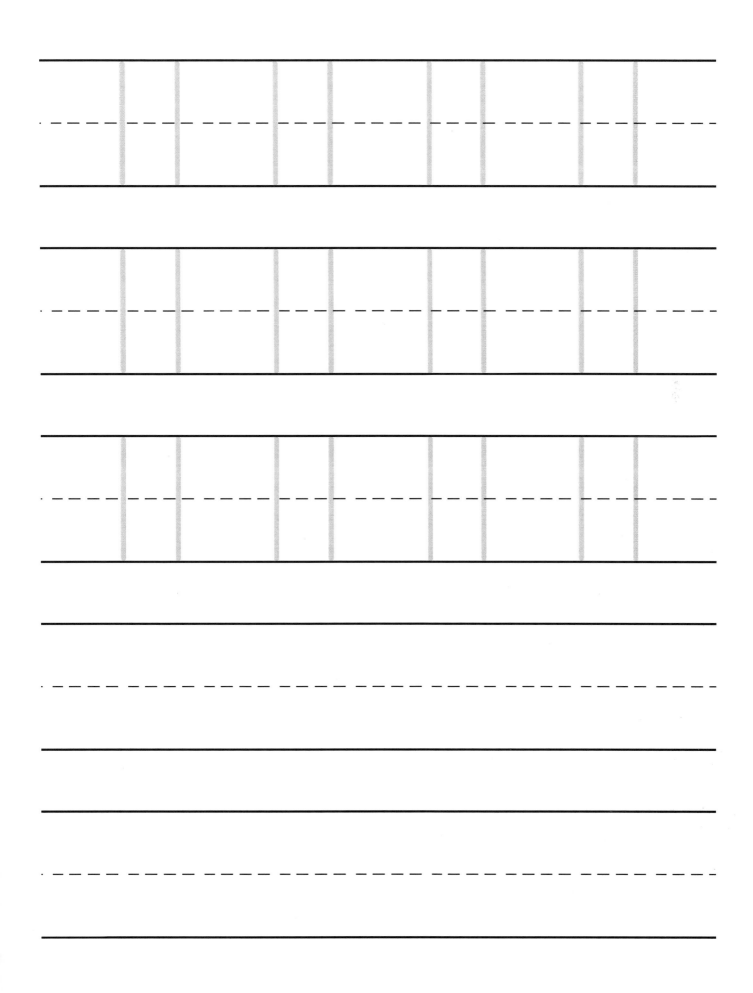

twelve

12 12 12 12

12 12 12 12

12 12 12 12

thirteen

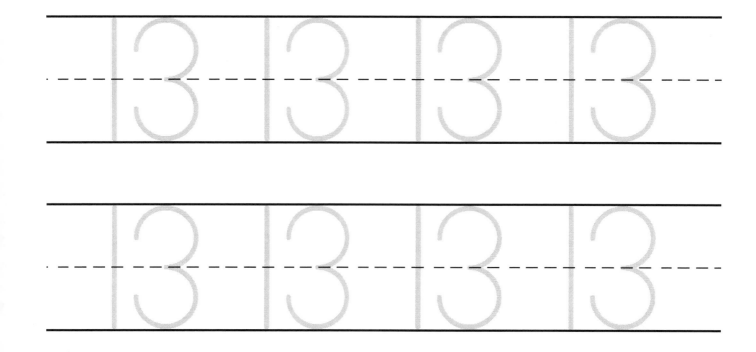

13 13 13 13

13 13 13 13

13 13 13 13

fourteen

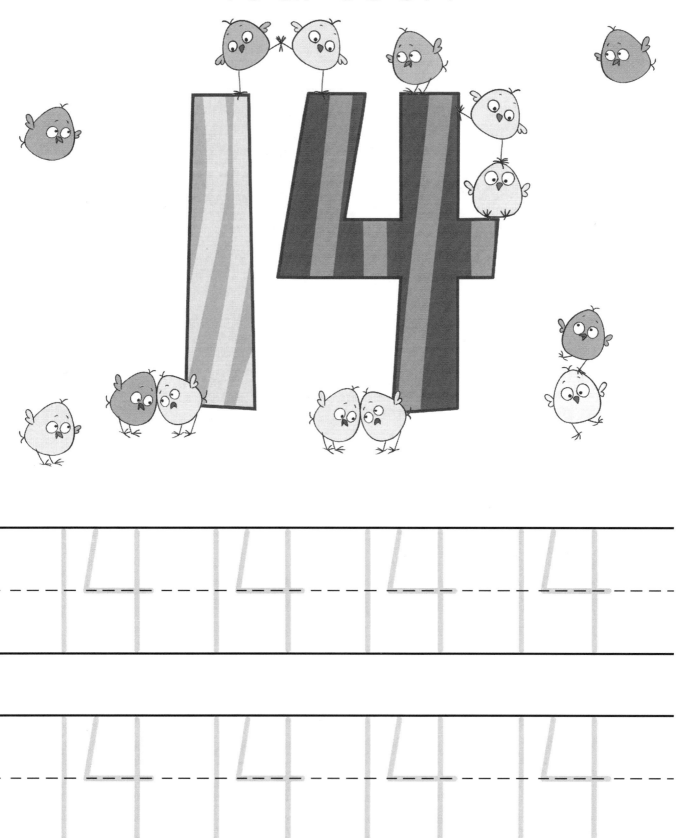

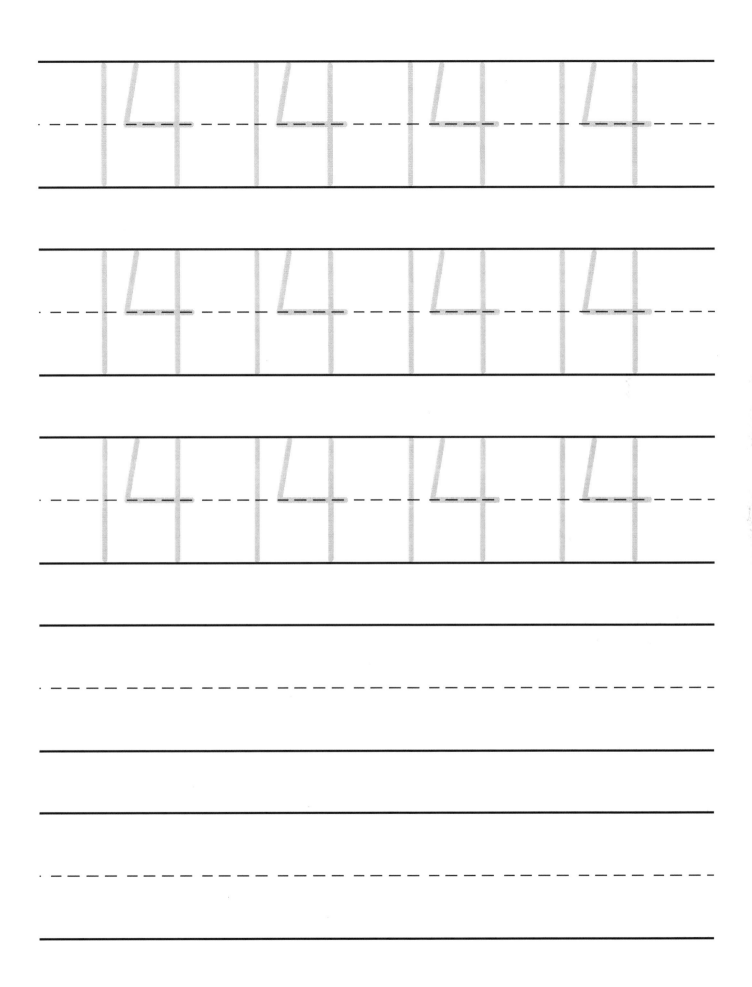

fifteen

15 15 15 15

15 15 15 15

15 15 15 15

sixteen

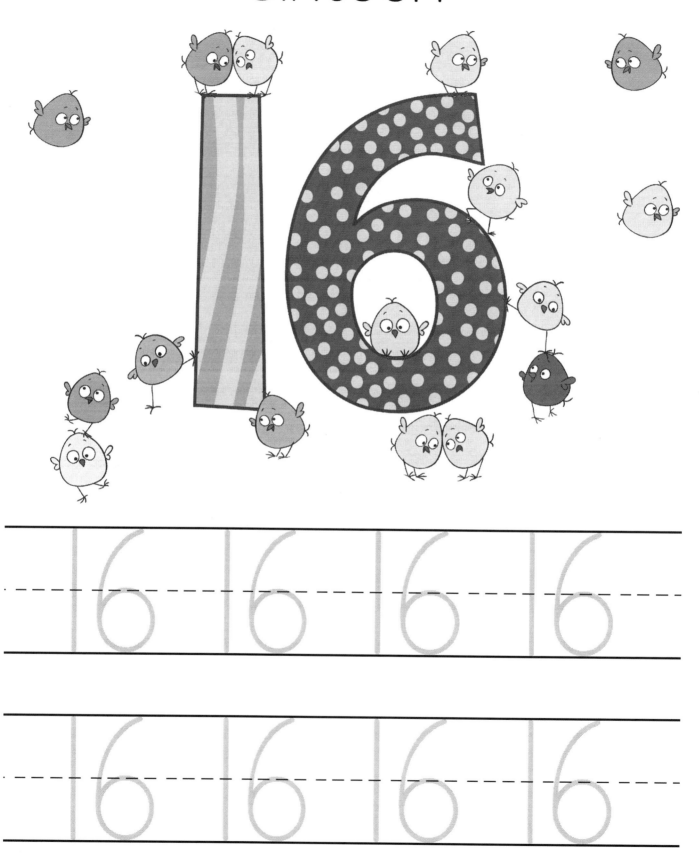

16 16 16 16

16 16 16 16

16 16 16 16

seventeen

eighteen

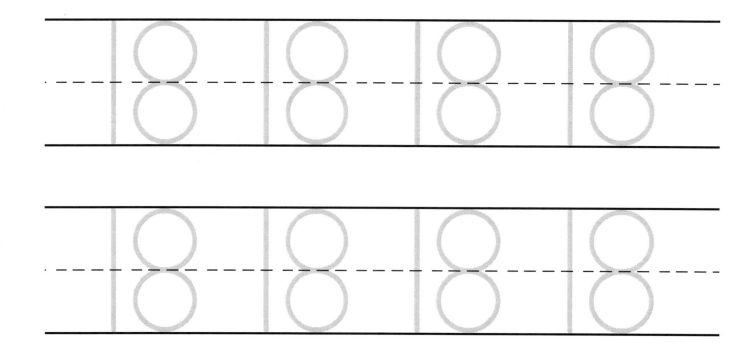

8 8 8 8

8 8 8 8

8 8 8 8

nineteen

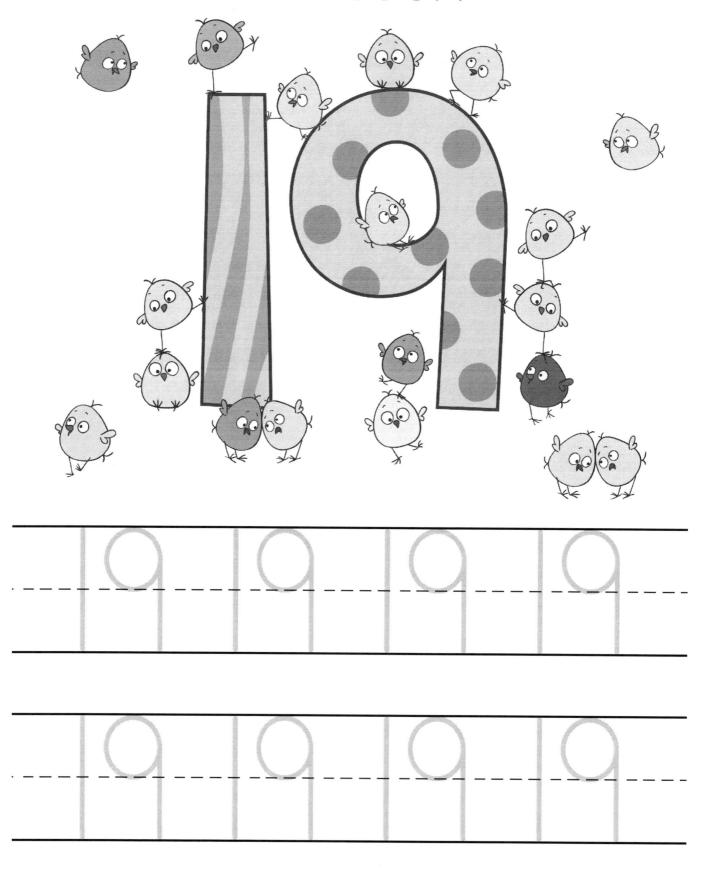

19 19 19 19

19 19 19 19

19 19 19 19

twenty

20 20 20 20 20

20 20 20 20 20

20 20 20 20 20

Part 2: Counting

Count the birds
- turn over for the answer

There was one bird

Count the birds
- turn over for the answer

There were two birds

Count the birds
- turn over for the answer

There were three birds

Count the birds
- turn over for the answer

There were four birds

Count the birds
- turn over for the answer

There were five birds

Count the birds
- turn over for the answer

There were six birds

6 6 6 6
6 6 6 6

Count the birds
- turn over for the answer

There were seven birds

Count the birds
- turn over for the answer

There were eight birds

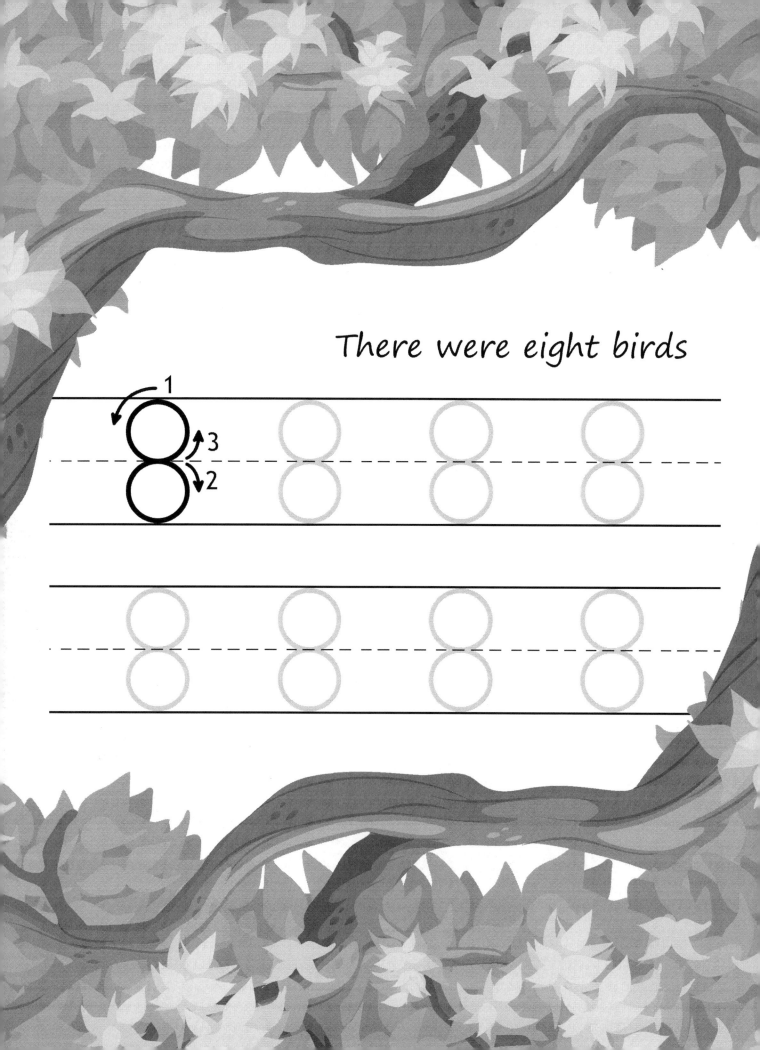

Count the birds

- turn over for the answer

There were nine birds

Count the birds
- turn over for the answer

There were ten birds